3 Dozen

JoAnn Frank

Sun Press, New York, New York
Printed by Michael Becotte and Chuck Gershwin in Phila., Pa.

Library of Congress Catalog Card Number 77-77575.
ISBN 0-9601260-1-5

Cover design by James Sherman and JoAnn Frank

To Jim Enyeart

"Meanwhile, down in the laboratory above the ice caves everything was very quiet except for the movement of a shadow. It was a shadow that just barely existed between forms. At times the shadow would almost become a form. The shadow would hover at the very edge of something definite and perhaps even recognizable but then the shadow would drift away into abstraction."

The Hawkline Monster: A Gothic Western by Richard Brautigan.
Published by Simon and Schuster, New York, N.Y. © 1974.
A Touchstone Book, p. 125.

egg - n. 1. the roundish reproductive body produced by the female of animals, consisting of an ovum and its envelopes, being albumen, jelly, membranes, egg case, or shell, according to species.

light - n. 1. that which makes things visible or affords illumination. 3. the sensation produced by stimulation of the organs of sight. 4. an illuminating agent or source, as the sun, a lamp, or a beacon.

shadow - n. 1. partial darkness or obscurity within a part of space from which rays from a source of light are cut off by an interposed opaque body. 2a. a reflected image. 4a. an imperfect and faint representation: an indistinct image. b. an unreal appearance or image: an imaginary or delusive vision. 6. phantom.

abstraction - n. 2a. the act or process of separating the inherent qualities or properties of something from the actual physical object or concept to which they belong.

exploration - n. 1. the act or an instance of exploring or investigating; examination. 2. the investigation of unknown regions.

photogram - n. . . . photography without a camera. The image of a shadow on a light-sensitive surface . . .

Notes

1. *The Random House Dictionary*, 1973 unabridged ed., s.v. "egg"
2. *Ibid.*, s.v. "light"
3. *Webster's Third New International Dictionary*, 1971 unabridged ed., s.v. "shadow"
4. *The American Heritage Dictionary*, 1976 New College Edition, s.v. "abstraction"
5. *The Random House Dictionary*, 1973 unabridged ed., s.v. "exploration"
6. *Moholy-Nagy, Marginal Notes: Documentary Absurdities* by Lucia Moholy. Published by Scherpe Verlag Krefeld, Germany ©1972, p. 59

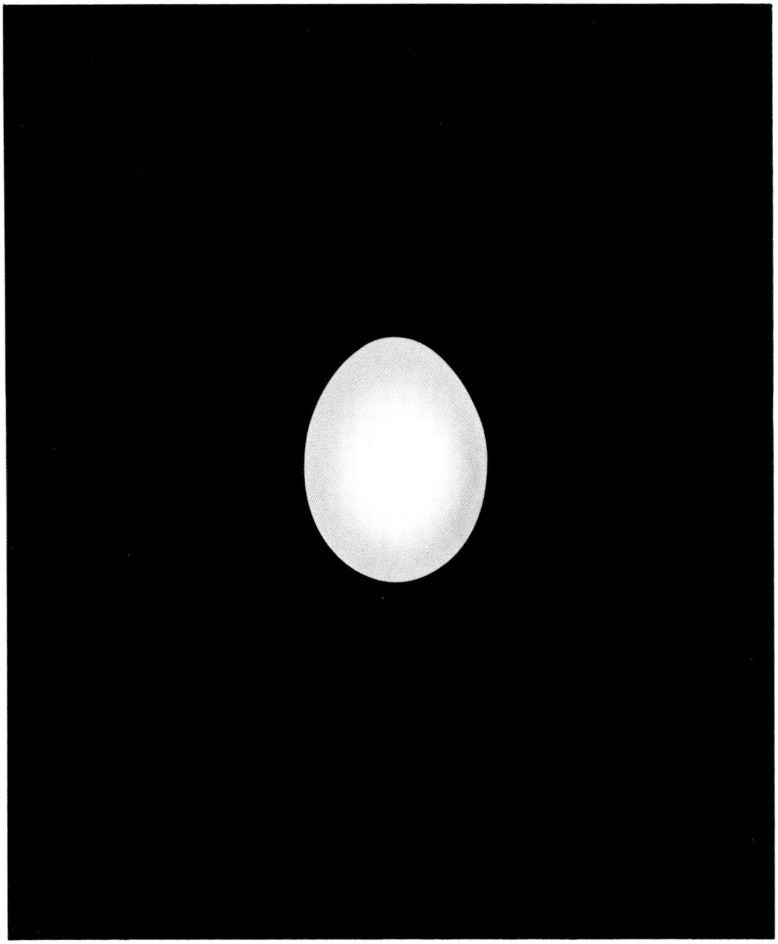

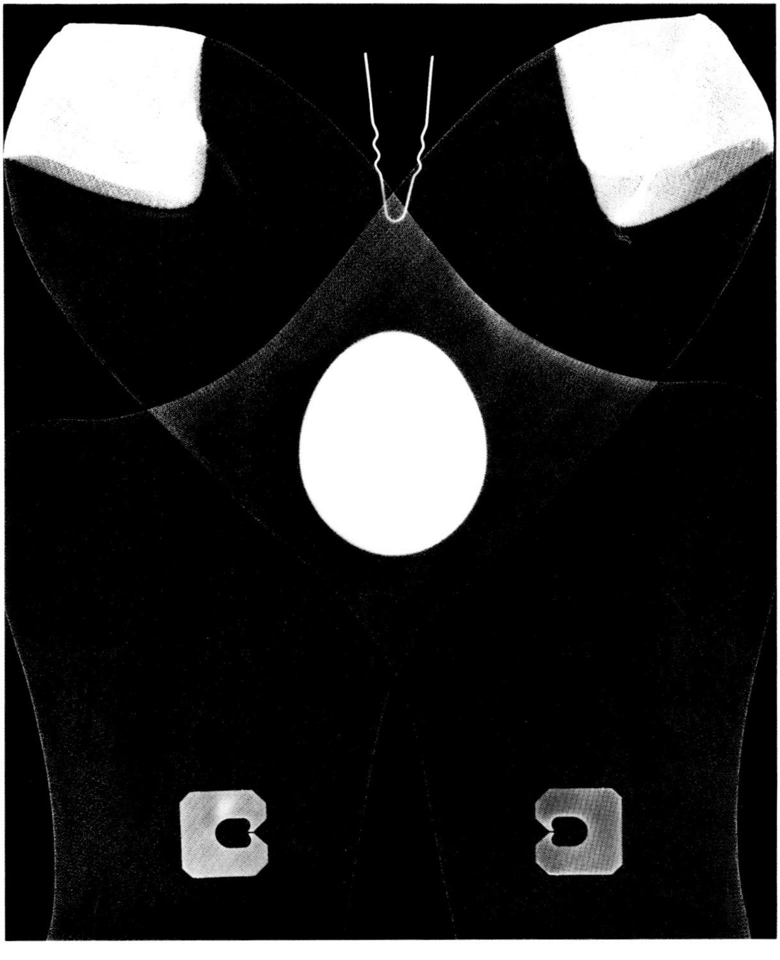

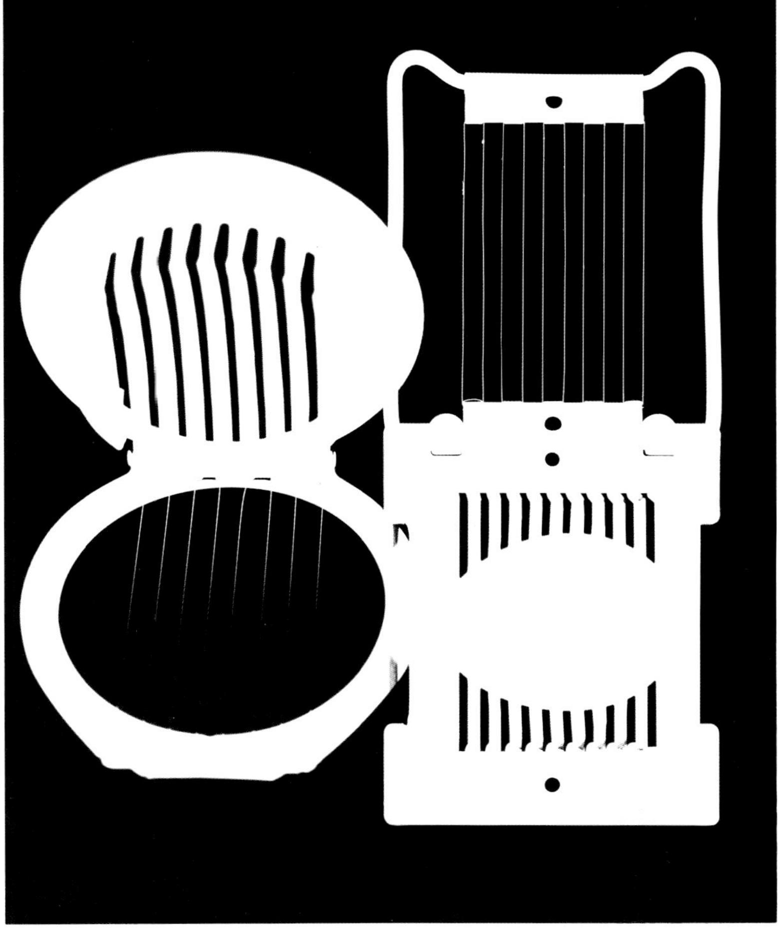

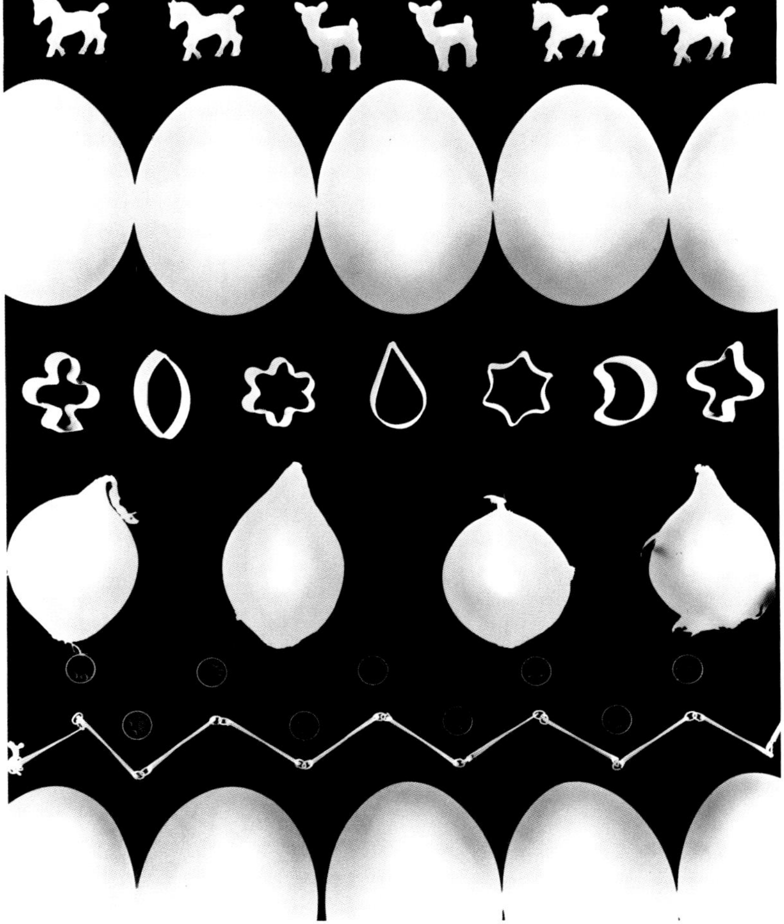

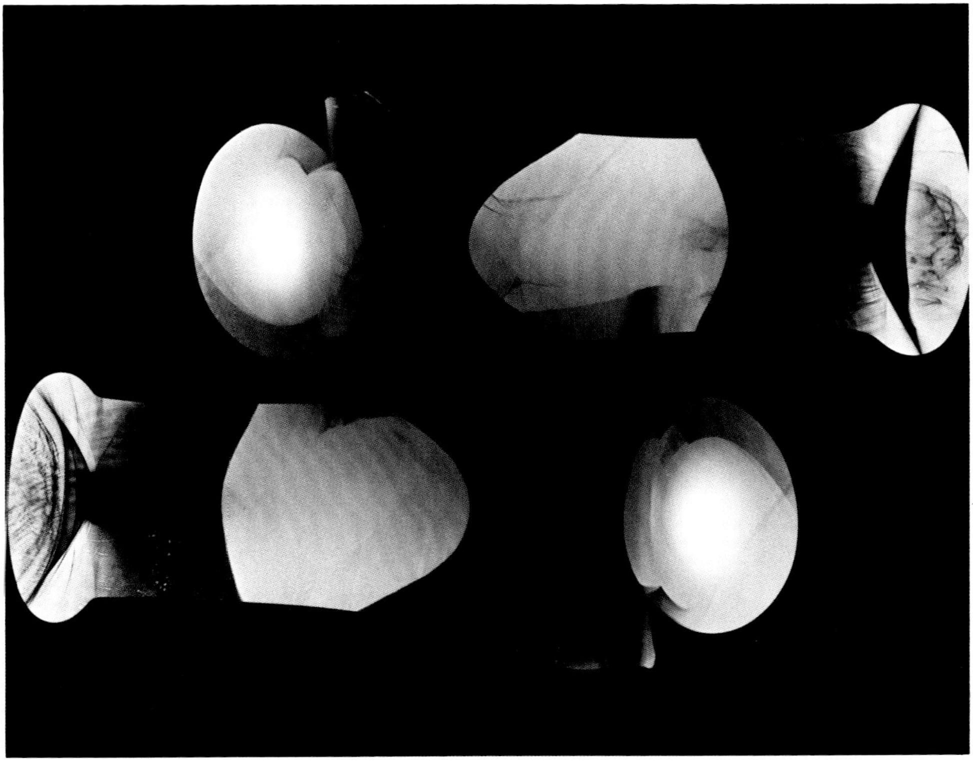

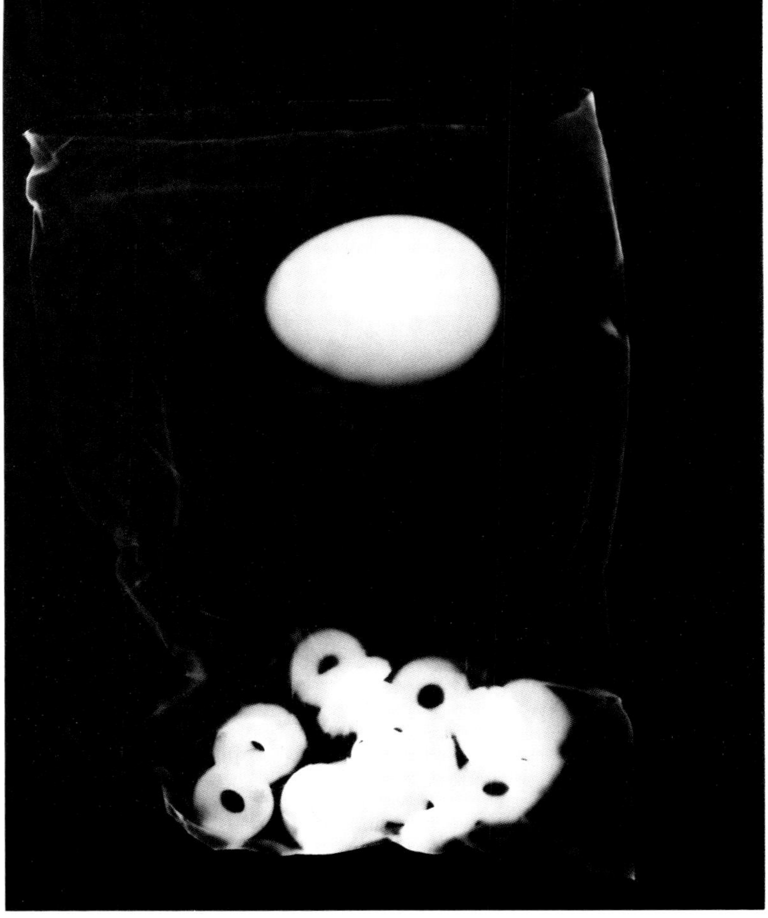

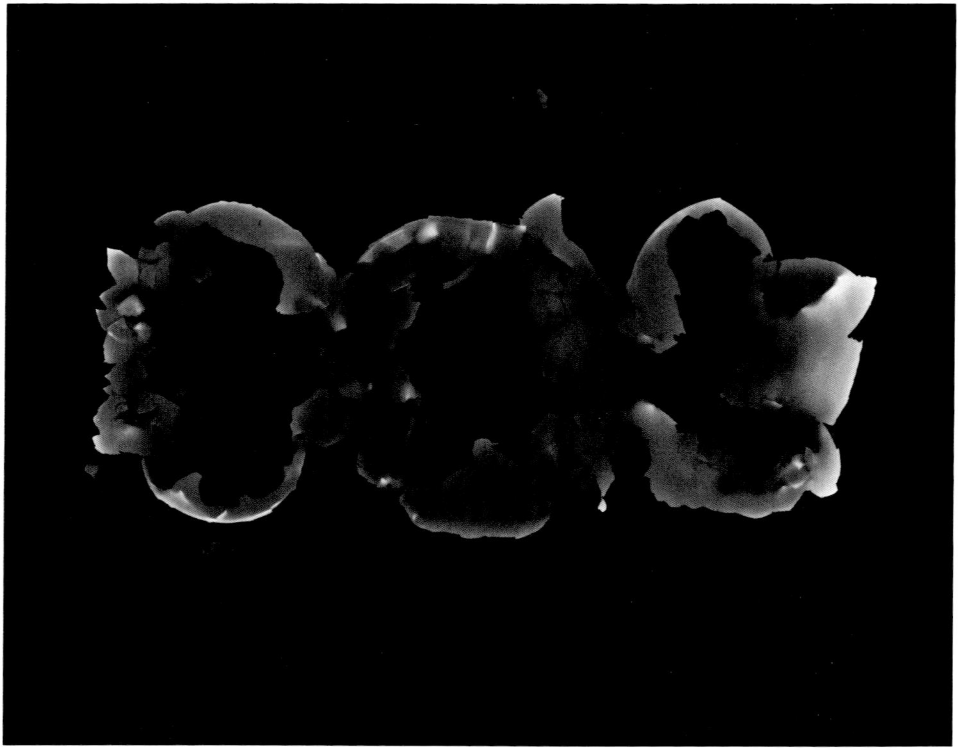

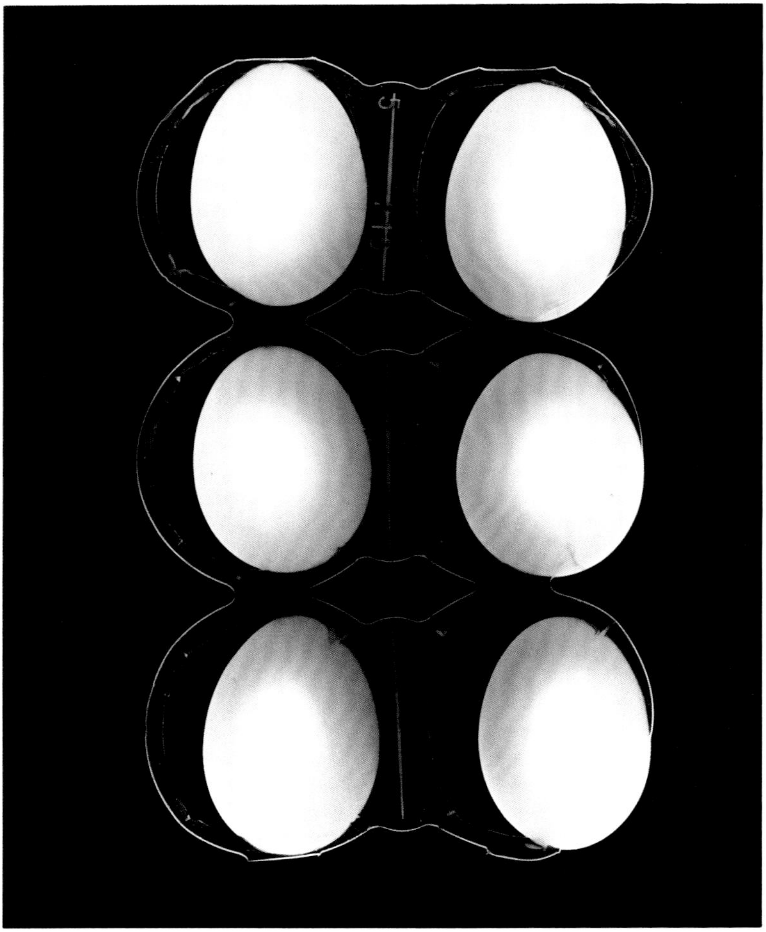

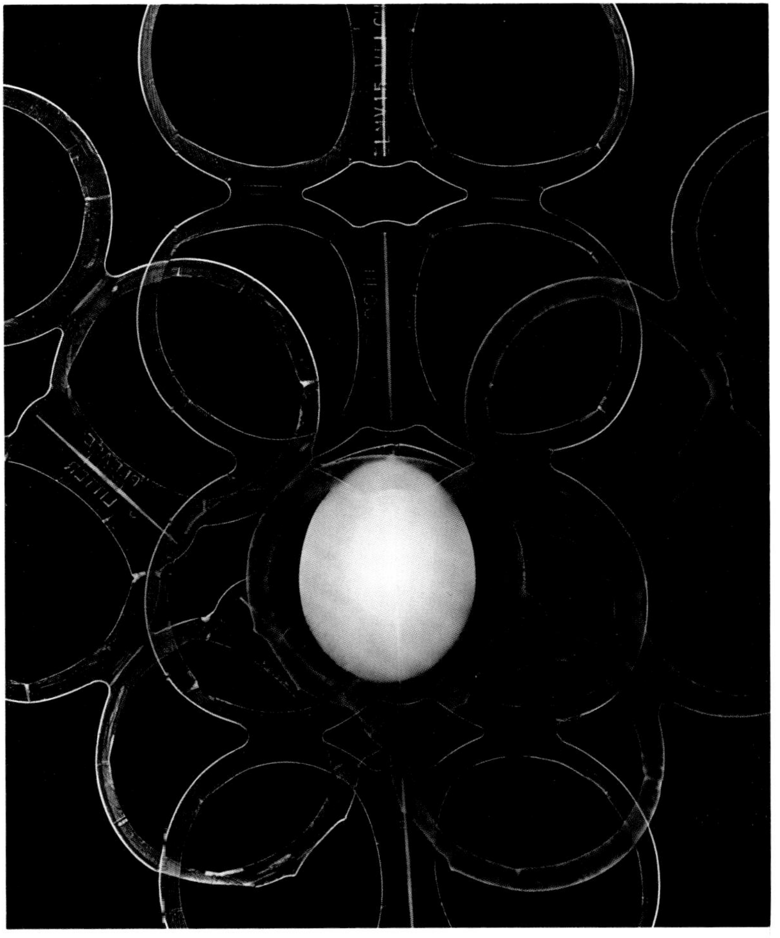

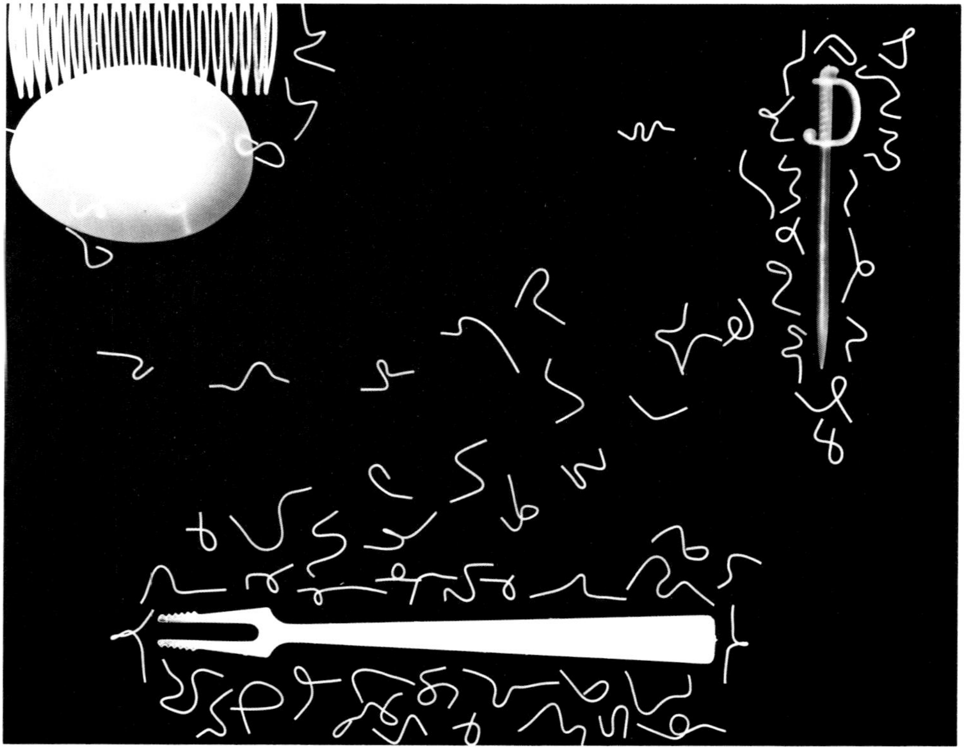

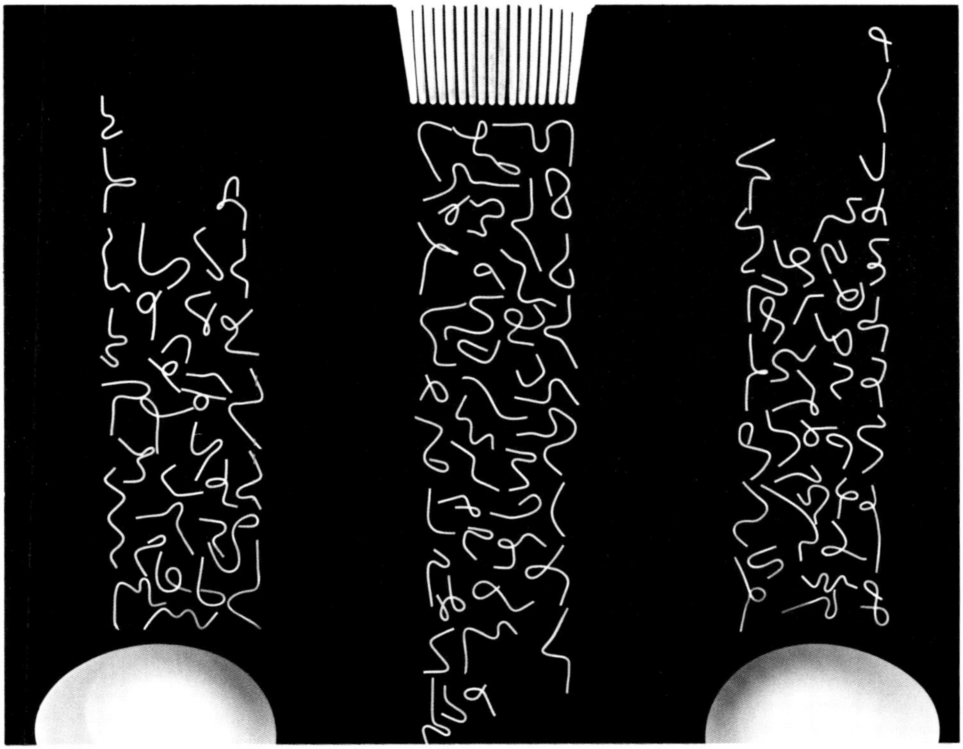

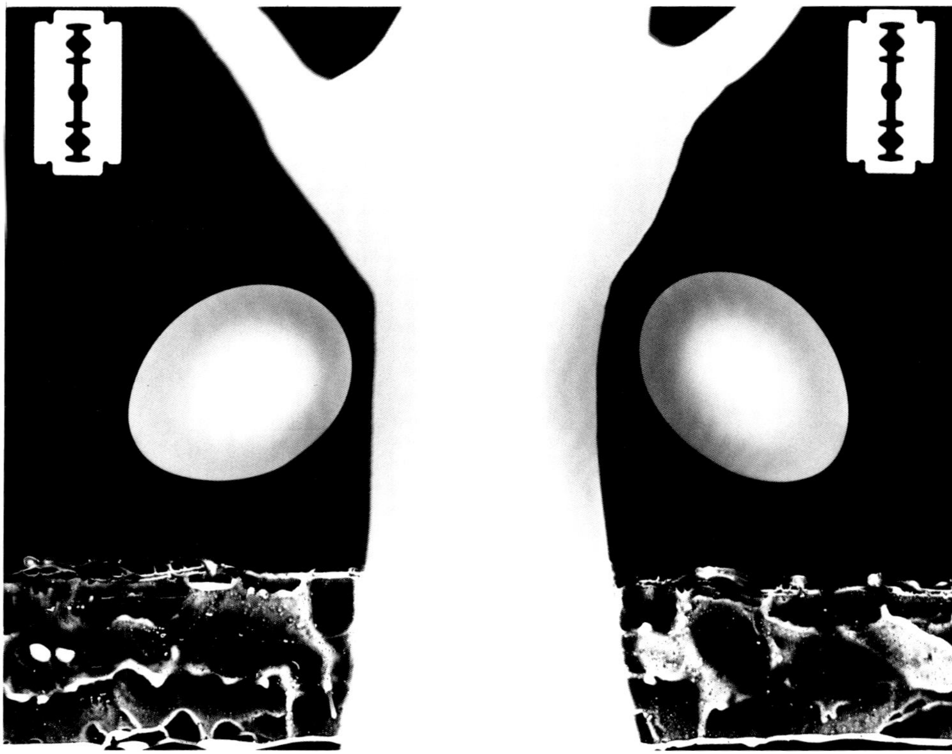

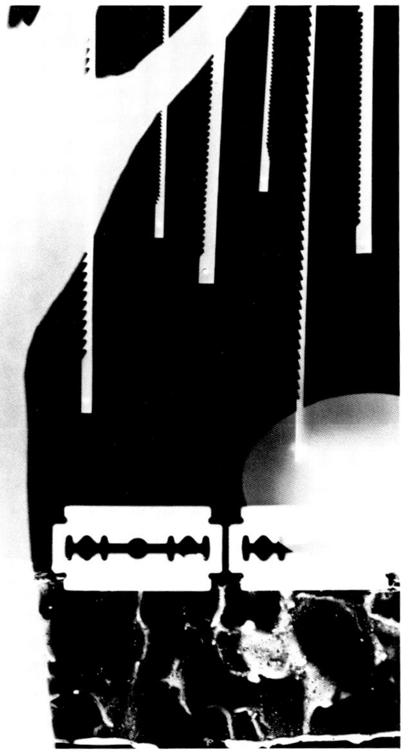

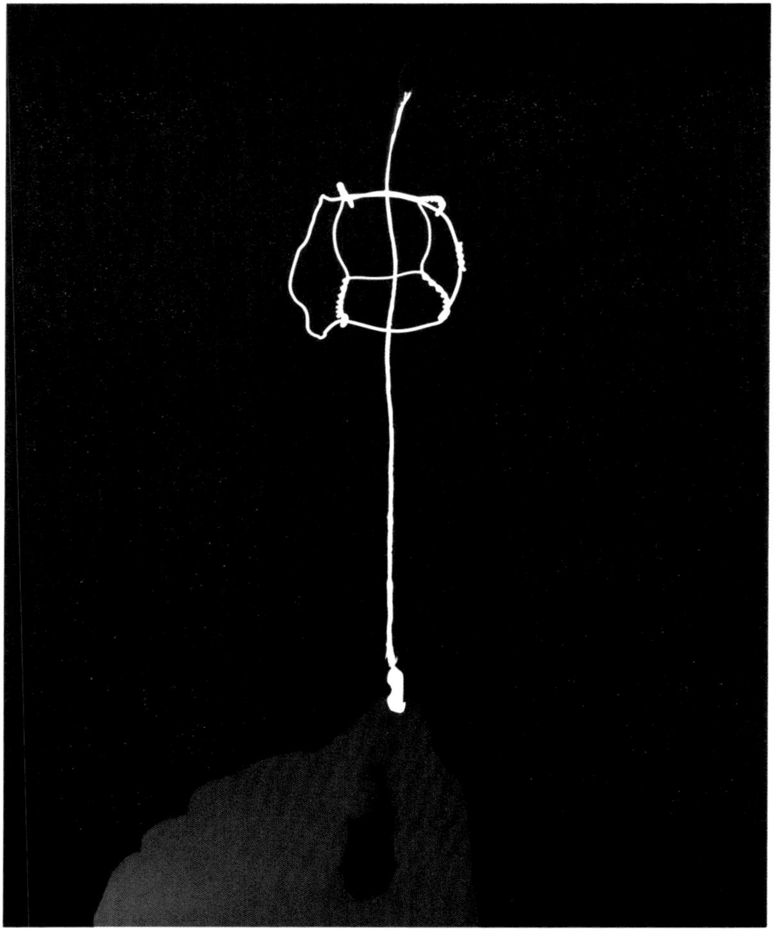

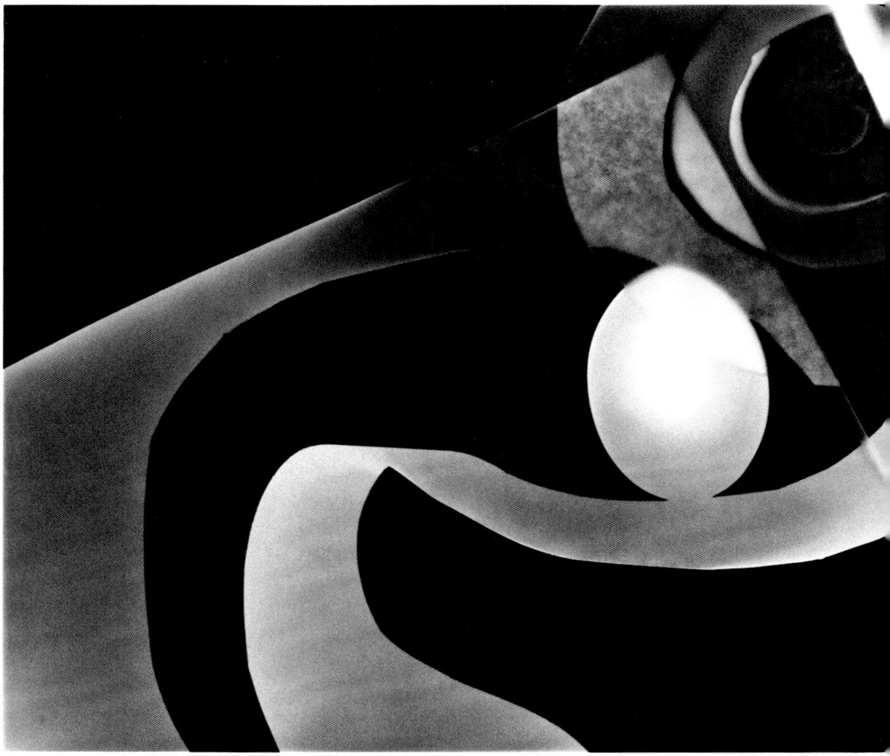

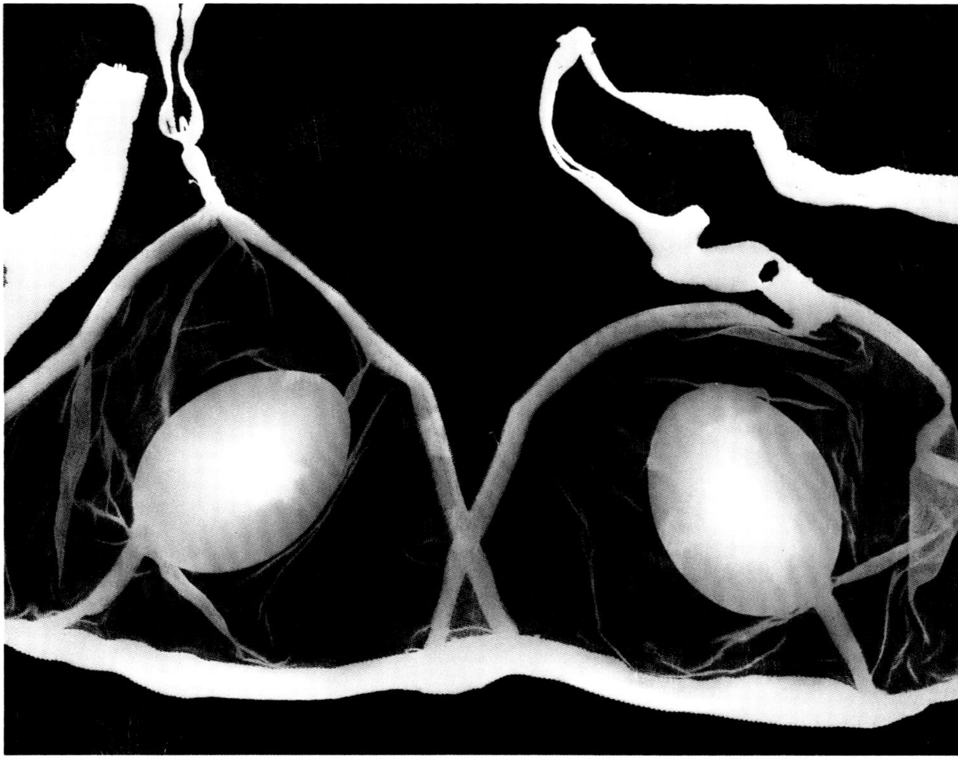

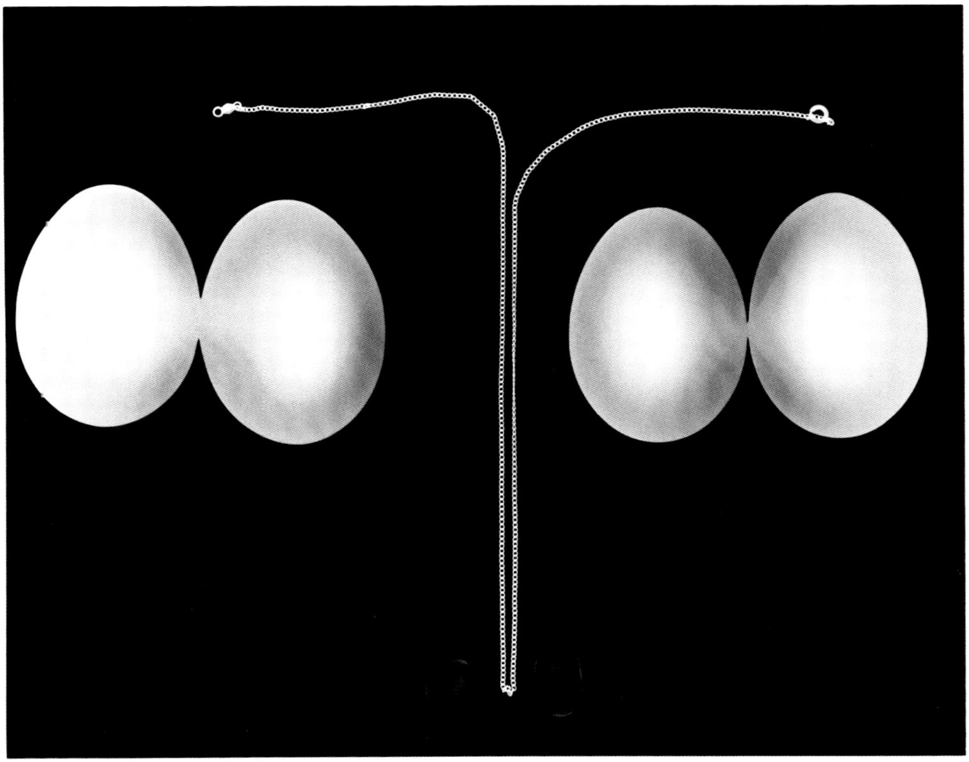

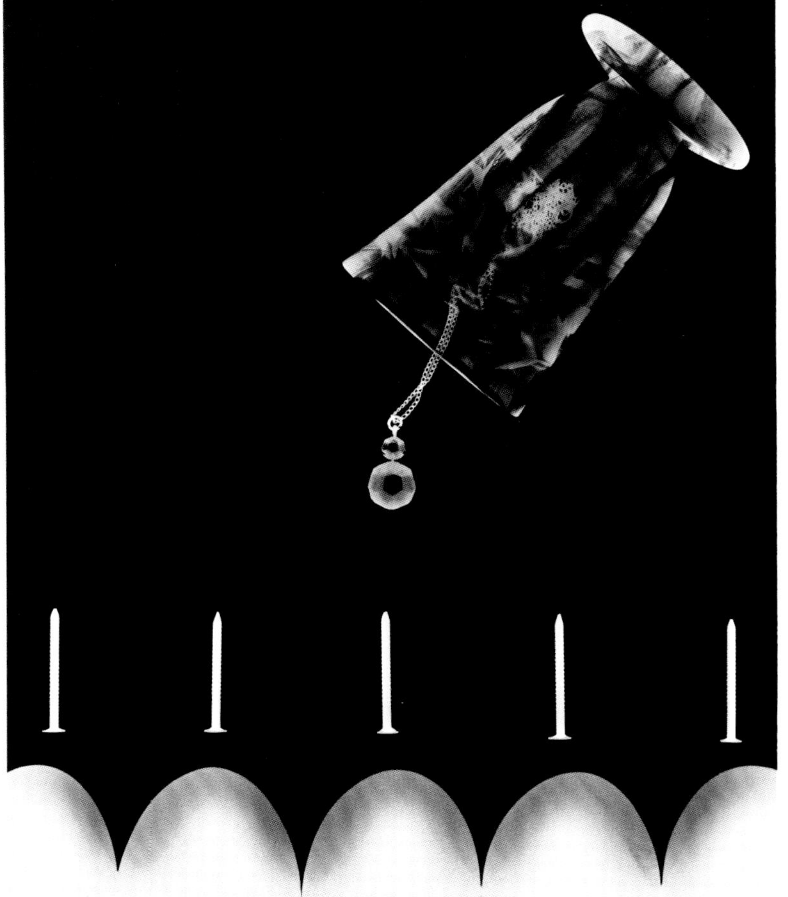

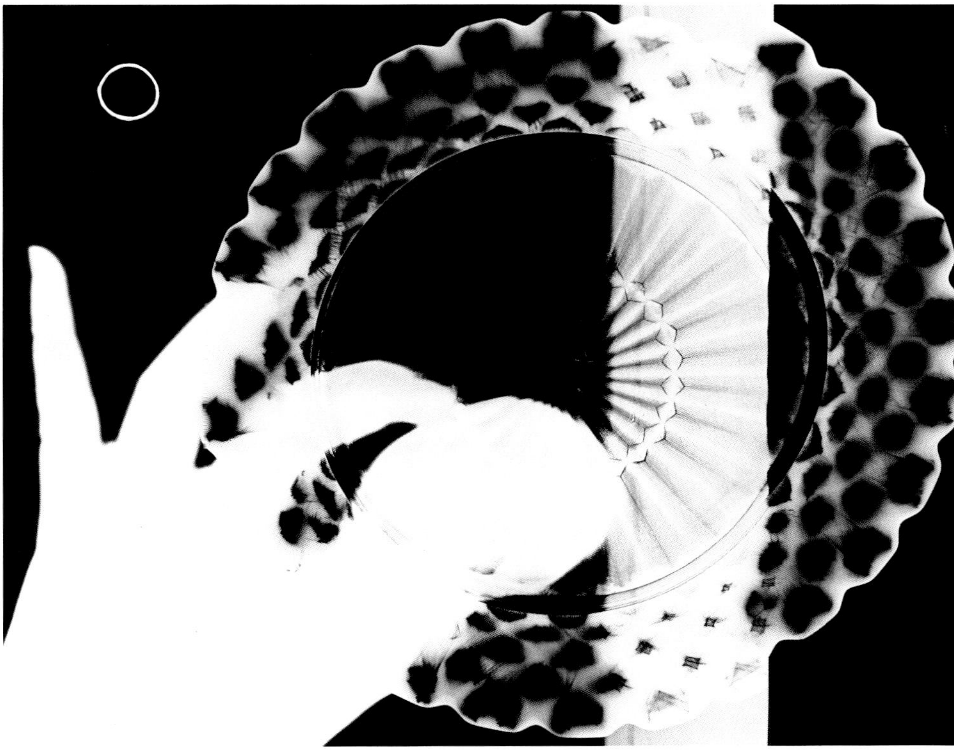

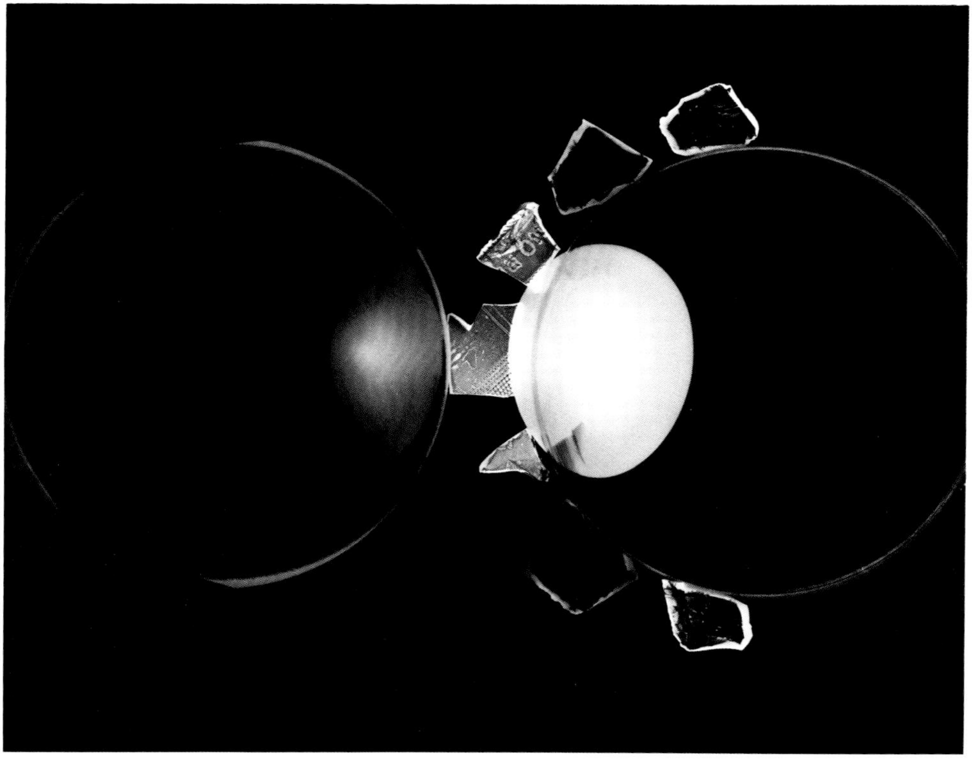

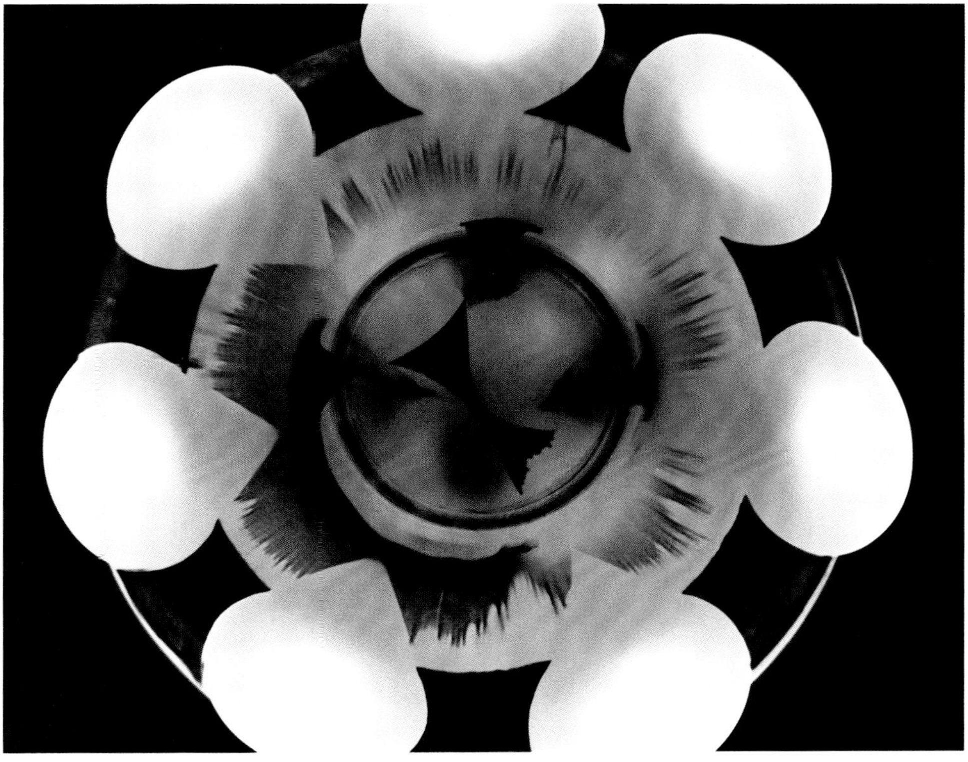

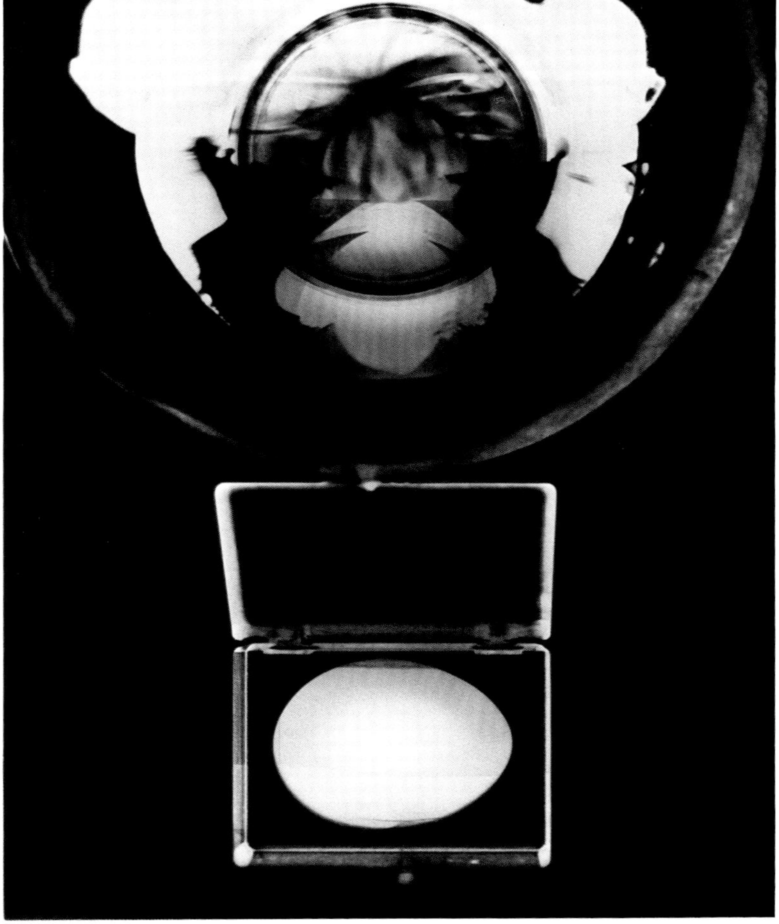

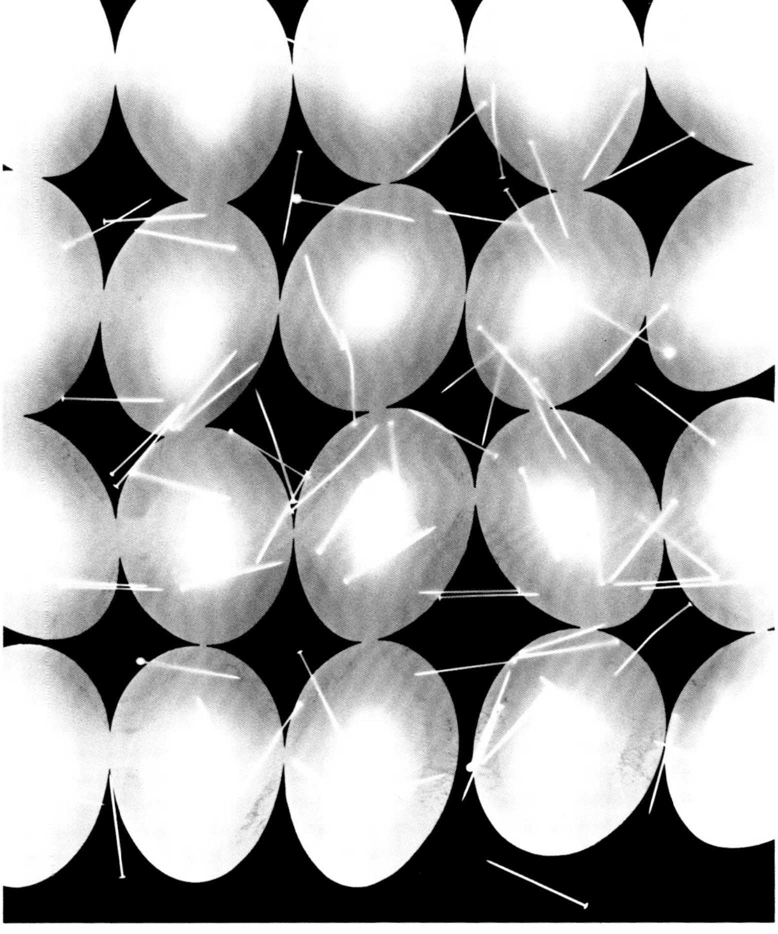

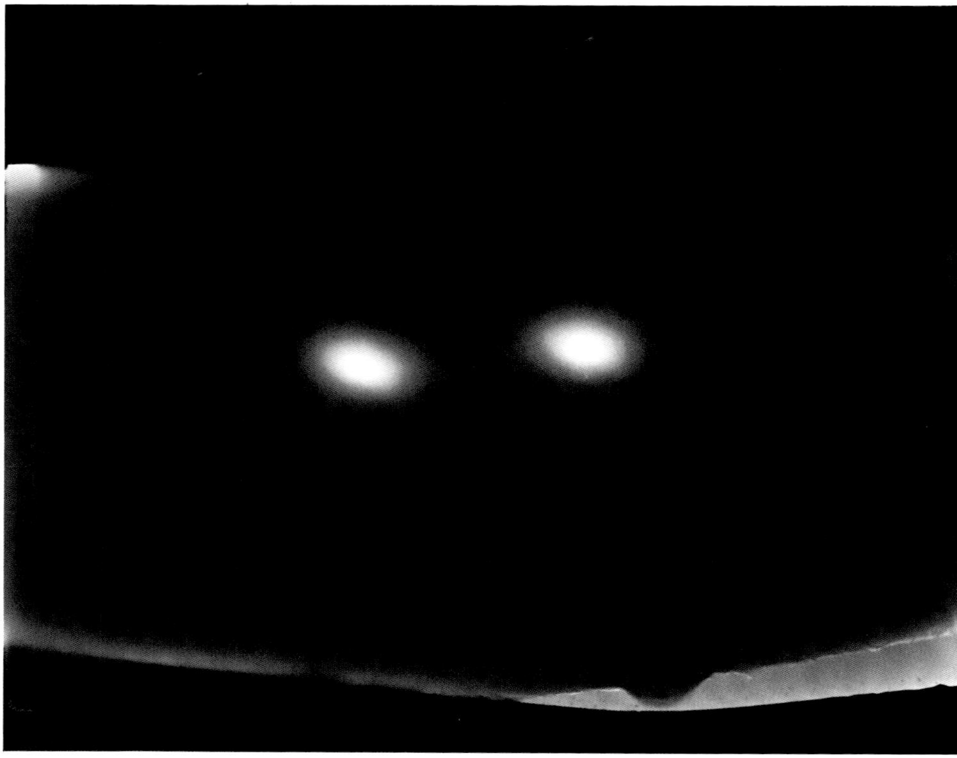

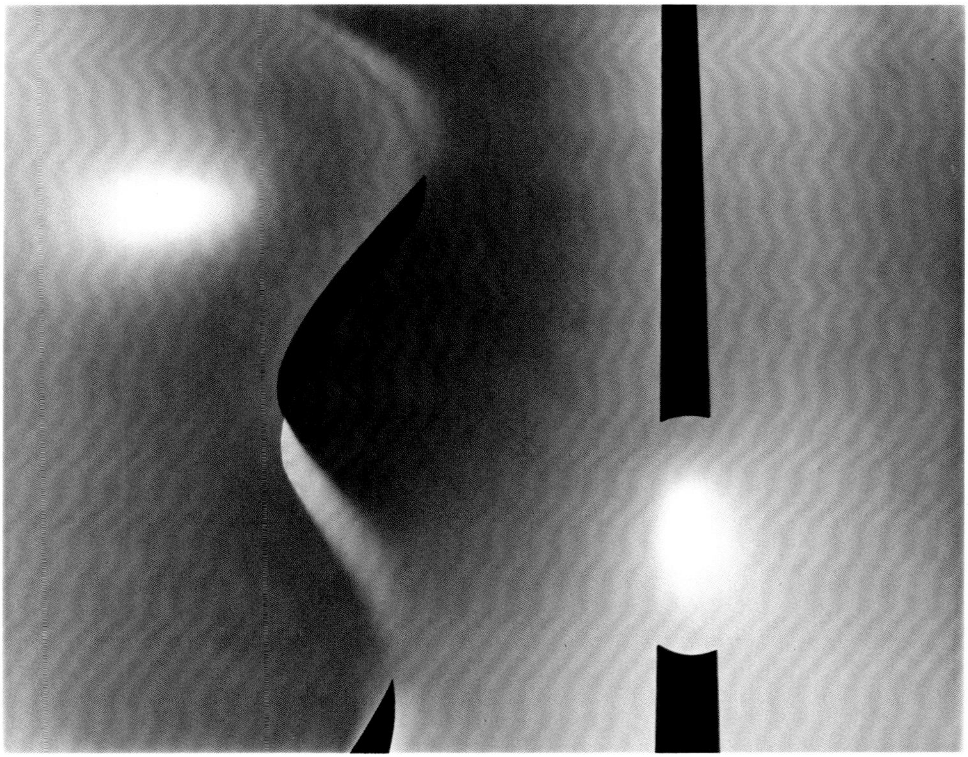